# THE MODIFIED GOLDEN HIVE

*Einraumbeute*

## DAVID HEAF

The Modified Golden Hive *(Einraumbeute)*
ISBN: 978-1-914934-24-7



Text and Graphics: David Heaf

Published by Northern Bee Books 2021
Northern Bee Books, Scout Bottom Farm,
Mythomroyd, Hebden Bridge, HX7 5JS (UK)
www.northernbeebooks.co.uk
Tel: +44 (0) 1422 882751

Designed by: Lynnette Busby
www.whatever.graphics

About the author

After retiring from research biochemistry to north-west Wales, David Heaf took up hobby beekeeping in 2003. He runs about 10 colonies in Warré hives and several in other hive types. In 2010, Northern Bee Books published his book *The Bee Friendly Beekeeper* and again invited him to write two other books, *Natural Beekeeping with the Warré Hive* – A Manual and *Treatment-Free Beekeeping,* (IBRA & NBB) which they published in 2013 and 2021 respectively.

# THE MODIFIED GOLDEN HIVE

*Einraumbeute*

DAVID HEAF

## Contents

## Introduction

The modified golden hive described in this book is an insulated double-walled trough or horizontal hive having up to 22 deep frames and a Warré hive style quilt and roof configuration. Its immediate precursor is the *Einraumbeute*, a German name which can be translated variously as one-box hive, or single chamber hive, etc.

Although my switch from UK National hives managed conventionally to more natural beekeeping predominantly involved the hive of Abbé Émile Warré,[1] I remained interested in trying the golden hive because it had been adopted by so many of my contacts who were also interested in more bee-friendly ways of keeping bees.

In 2014, before embarking on my modified golden hive project, I gained some experience with the horizontal hive format by making a modified version of the hive of Fedor Lazutin (1966-2015).[2] It too was given double walls and a Warré style top-of-hive assembly. I designed that hive around UK National hive top-bars which I incorporated in 450 mm deep frames, and for the hive body and roof used various pieces of suitable scrap wood that I could find, mostly from a plant processing municipal wood waste destined for incineration in a power station.

If I was to do the same thing with a modified golden hive, I first had to secure a set of 22 frames around which to build the hive body. Fortuitously also in 2014, Matt Somerville and John Haverson made a modified golden hive,[3] and as Matt, a professional woodworker, was tooled-up to make the frames, a job requiring much precision, he kindly made me a set to order.

## Naming the Hive

The Mellifera *Einraumbeute* (ERB),[4] was first conceived in 2000 by Thomas Radetzki of Mellifera Association, Fischermühle, Germany,[5] where some 20 examples of it were populated in 2002.[6] Its frames are Dadant size rotated through 90°, i.e. from landscape format to portrait format. It has been named the 'golden hive' because its frame and shape are based on the golden section/ratio/mean/proportion. But we might question this naming, on the grounds firstly that the Dadant hive precedes the *Einraumbeute*, and secondly because the latter's inventor decided in 2003 no longer to call it the 'golden hive', and requested that it should be simply called the 'Einraumbeute' or if necessary the '*Einraumbeute*' of Mellifera e.V.[7]

For its naming in English, directly translating its German name has so far been avoided. As the brood box of a hive is often referred to as the 'brood chamber', one possible English name would be the 'unicameral hive'. Another possibility is 'one-box hive'. But in anglophone beekeeping circles a common name for this hive type is the 'horizontal hive' because the space occupied by the colony is mostly expanded horizontally instead of vertically, as in the case of the Langstroth hive (USA) or the National hive (UK). The currency of the term 'horizontal hive' applied to frame hives has increased greatly due to the work of Leonid Sharashkin in publishing in English translation the books by Fedor Lazutin and Georges de Layens about their versions of horizontal hives.[8] A 14-frame Layens with 32 mm thick walls in UK cedar is available from Thorne who report on their web site that "this hive is becoming more and more popular with thousands now in use in the USA and over 1 million in Spain alone".[9]

Before leaving this issue of naming the hive, we should not forget that a hive of similar format was introduced in the UK by Robin Dartington.[10] His 'long, deep hive' is based on National frames sized 14" by 12" and also has a horizontally expanded brood nest. It is designed to take small National size supers. His whole hive concept, including its rotateable legs that double as handles for carrying, rather in the manner of the sedan chair, is designed to minimise lifting of heavy weights. The Dartington hive format was incorporated in the 'Beehaus' a plastic cased hive.[11] A similar concept appears in the Drayton hive with its eighteen 14" x 12" frames housed in a well insulated hive body with a quilt and ventilated roof.[12]

Last but not least, there is the Zest hive, a horizontal hive designed by Bill Summers. Its walls are made from aerated concrete blocks and it contains plastic frames.[13] It has an insulated top cover. Its walls alone must make it more than a match for the thermal performance of a tree cavity!

## The Golden Hive's Precursors

Long deep trough or horizontal hives were relatively common in eastern Europe including Russia. For example, Fedor Lazutin describes a Ukrainian hive with 24 Dadant-sized frames turned through 90° and measuring 285mm wide by 460mm deep (Fig. 1).[14] Ferdinand Gerstung, in Germany, also presented a hive with a deep format called the *Thüringer Einbeute* whose frames were 260 mm wide by 410 mm deep.

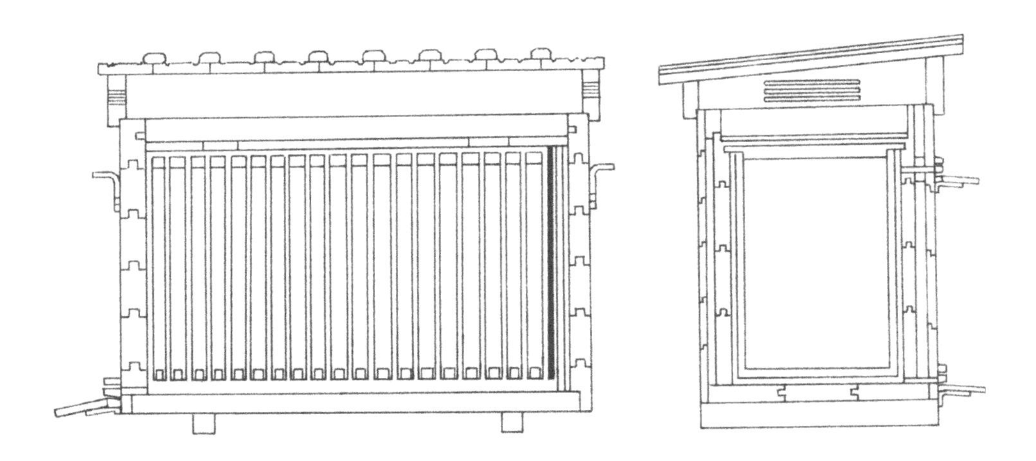

*Fig. 1. Ukrainian hive cross sections (side and end views)*[14]

## Advantages of the Golden Hive Format

In their book *Keeping bees simply and respectfully -- Apiculture with the Golden Hive,* Johannes Wirz and Norbert Poeplau present the advantages of the hive as follows:

- The bee colony develops in a single chamber and is not divided;
- The combs are constructed naturally and there is usually sufficient honey for the bees' consumption through summer;
- The brood and honey zones can be easily extended without disturbing the colony;
- When inspecting the brood and/or honey zones, no heavy supers have to be lifted off;
- The working height can be adjusted to the individual at initial installation, thus enabling back-friendly working.[15]

To those points we can add that the insulated double wall and Warré hive style quilt of the modified golden hive protect against extremes of heat. Furthermore, the roof, with its well ventilated cavity underneath, a feature also borrowed from the Warré hive, shields against the sun. These modifications also add to the thermal mass of the hive, further helping to buffer temperature fluctuations and thereby making it easier for the colony inside to control its temperature. Compared with the thin-walled hives in common use, the modifications also bring the hive's thermal performance somewhat closer to that of a tree cavity.

## Construction of the modified golden hive

To sum up the main design features:

- Double walls with insulation in between (inspired by Fedor Lazutin's trough hive);
- Warré hive top-bar hessian cloth, quilt and roof configuration;
- Entrance at one end through a notch in the floor (similar to Warré hive floor), resulting in 'warm way' comb orientation with respect to the entrance;
- No mesh floor or Varroa tray, but a solid floor that can be slid out for cleaning;
- Provision for emergency ventilation in front, rear and floor;
- Legs fixed to the hive and of a length to give a comfortable working height.

Apart from adhering to those features and making the inside dimensions fit 22 frames, allowing for bee space, I did not use construction plans. However, those wishing to make a modified golden hive and who are confident enough to proceed without plans may find sufficient information from the description and photos in this book. Whereas those who would feel more comfortable embarking on such a project with plans to refer to can use the Haverson/Somerville plans that are freely downloadable.[15] Furthermore, for those who wish to add their own modifications to the original golden hive, plans for the authentic *Einraumbeute* are also available in German for free download from Mellifera's website.[16]

Before starting to make the hive, it is advisable to secure a set of frames and be sure of their dimensions.

*Fig. 2. Inner box*

*Fig. 3. Inner box with flanges added*

I used the frames as a template when building the inner walls of my modified golden hive, allowing an 8 mm bee space at the sides and bottom. The wood happened to be planed 20 mm pine, probably originally serving as bed slats. The fixings were mostly nails and exterior glue except where stated otherwise. The top and bottom flanges served to set the width of the cavity in the walls, and the thicker top flange to take a rebate for the frame lugs. (The hive was inadvertently photographed inverted for Fig. 3.)

*Fig. 4. Fitting cladding and insulation*

*Fig. 5. Fitting cladding and insulation*

The cladding was 20 mm rough sawn softwood. Its planks were fixed to the flanges of the inner box with corrosion-resistant screws. While fitting the cladding, the cavity was packed with dry *Phragmites australis* (reed) straw harvested from a nearby estuary. The straw was installed to stand vertically in the finished hive to eliminate the possibility of the filling settling, which would otherwise remove some of the insulating benefit. Alternatively, sufficiently long rye straw could be used or any natural insulating material not inclined to settle, e.g. cork board, sheep wool. Over the seams between the abutting cladding planks were nailed thinner strips of wood to minimise water penetration.

*Fig. 6. Cladding completed; vents shown in cladding and floor*

*Fig. 7. Floor partly slid out*

Before proceeding with the description of hive construction, a note about the sustainability of this project is in order: by comparison with most hives, making a double-walled hive may seem like an extravagant use of wood, especially in these times when climate change mitigation measures such as tree planting are in the news. But as the wood I used was destined for large-scale incineration, my use of wood for this project had even less environmental impact than making single-walled hives from new wood. Of course, not everyone will be able to intercept reusable wood destined for the waste stream, but it is worth making enquiries locally to see if there are any wood re-use businesses nearby.

Emergency vents were drilled in the walls and floor. For the vents in the walls, the gap between the inner and outer walls was bridged by plastic pipe. When the vents are not in use, the holes are plugged with wine corks. More recent models of Mellifera's Einraumbeute also include vents.[17] However, it should be noted that in many seasons use of my modified golden hive, I never found it necessary to open the vents.

Using stainless steel screws, the four legs are screwed onto the long sides of the hive, opposite the inner vertical members of the wall frame at each end of the inner box. The leg length was chosen to give a working height that suited myself. The wood used in the legs, having a greenish tinge, looked like softwood that had been pressure-treated against rot.

*Fig. 8. Entrance detail*

*Fig. 9. Entrance detail with floor slid forward*

In my design, the legs play an important part in supporting the hive's floor. Referring to figures 8, 10 and 14: firstly two rails of approximately 50 x 25 mm section softwood were installed, one between each pair of front or back legs. On these rest two similar rails running between the front and the back of the hive (#5 in Fig. 8). On these rest the floor made of the same wood as the cladding (#3 in Fig.8). To keep the floor properly aligned with the underside of the trough, two guide battens of approximately 15 x 15 mm section (#4 in Fig. 8) are nailed to the long rails supporting the floor. Sufficient clearance of about 5 mm is allowed between the floor support rail (#5) and the bottom of the hive body, in order to allow the floor to be slid out from under the hive for cleaning. At all other times, the floor is clamped to the underside of the hive with wooden wedges ('w' in Fig. 10) inserted between the floor boards (#3) and the support rail (#5) at the front and back of the hive.

The first version of Mellifera Association's *Einraumbeute* had the entrance in the middle of the long side. This created elongated brood nests and difficulty when harvesting honey.[18] Their version current at the time of writing this book has three closeable entrances along one long side. Commonly one of the end entrances is used and some users have even moved the entrance to one of the short sides of the hive, as is the case with the modified golden hive presented here. It results in a single harvestable honey storage area at the back of the hive furthest from the entrance.

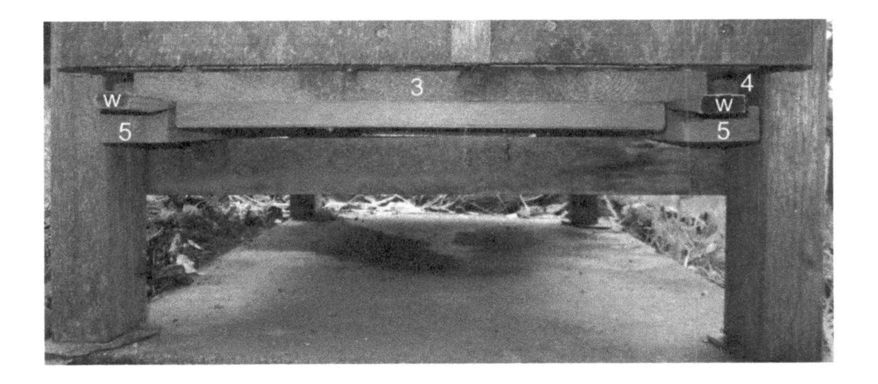


*Fig. 10. Rear view of floor support and wedges*

During the hive's subsequent use, had it been necessary to remove the floor frequently for cleaning, hinging one side to allow the other side to be dropped down for access would probably have been the better solution for installing the floor. As it turned out, the first colony to occupy the hive, and survived untreated for Varroa for 5 years, kept the floor remarkably clean.

Mellifera's *Einraumbeute* has a mesh floor and drawer for sampling and counting Varroa to assess the colony's burden of mites. As neither I, nor my contacts who use the modified golden hive, treat against Varroa or count mites, we do without a mesh floor.

*Fig. 11. Frames inserted, hessian top-bar cloth, quilts in background*

The top bars of the frames are covered with a cloth (Fig. 11.). Mellifera's *Einraumbeute* has an organic cotton cloth soaked in beeswax. As my Warré hives have functioned well for over 14 years with hessian (burlap) top-bar cloths, as recommended by Émile Warré himself,[19] it seemed appropriate to use the same with the modified golden hive. Jute sacking (hessian) can be obtained from pet shops selling peanuts for birds. Warré suggested treating it with a paste of starch and rye flour. This stiffens the fabric when dry and makes it less prone to being chewed by the bees. When inspecting the combs, it can be gently peeled back to the desired extent without the cracking that is associated with removing inner covers on some more conventional hives.

*Fig. 12. Quilt boxes in place*

Continuing with the Warré-style top-of-hive assembly: the modified golden hive has a 100 mm deep insulating layer on the top bar cloth comprising two boxes containing some kind of natural insulating material, in this case electric planer shavings which would otherwise be discarded. The quilt contents are retained by a piece of untreated hessian, and the quilt boxes covered with any kind of cloth to prevent the contents from blowing about in windy conditions when the hive is opened. The single quilt box of the Warré hive is doubled on the golden hive because a single box was considered unwieldy.

*Fig. 13. Roof woodwork*

The roof, also inspired by Warré's design, comprises fascia boards round its eaves, an internal 'mouse board' to prevent rodents getting into the quilt and a cavity under the roof covering that is ventilated at the eaves on both long sides (Fig. 13.). In order to prevent moisture ingress, especially in driving rain, it is very important that the fascia boards extend to at least 20 mm below the top-bar cloth. The waterproof covering of the roof can be any rigid, lightweight material, e.g. metal or plastic, or even thin marine plywood that is painted. In this example, scrap aluminium mobile home cladding was used, folded at the edges to increase rigidity. It is fixed to the woodwork with stainless steel screws.

*Fig. 14. Finished modified golden hive*

The finished hive (Fig. 14) stands on a firm base to prevent it sinking into soft ground. A 900 x 600 mm concrete slab was obtained from the local Freecycle network. Any necessary levelling of the hive after the slab is positioned can be done with pieces of slate waste. For convenience of viewing what the bees throw out of the entrance from time to time, a second recycled slab is placed in front of it.

The deep or 'portrait' frame format provides for a vertically undivided brood nest with a deep honey 'crown' very close to it. This format is uncommon in the UK. However, it is not recommended that beekeepers make the frames themselves, as it requires precision woodwork and, if carried out with a circular saw by amateurs, is notoriously costly in terms of beekeepers' fingers. As mentioned above, mine were kindly supplied by Matt Somerville who at the time of writing this book is willing to take orders for them.[20] Furthermore, since I made my golden hive, several sellers of *Einraumbeute* equipment including frames have appeared in Germany and neighbouring countries. At the time of writing they include sellers with the following web domains: holtermann-shop.de, fribin.de, mellifera.de, bienen-janisch.at, bienen-ruck.de, honigmobil.de and bijenhuis.nl. Some of these suppliers also offer spacers, follower boards and even transport/harvesting boxes.

*Fig. 15. Frame lower bamboo spacer*

*Fig. 16. Frame upper bamboo spacer*

My frames came without spacers of any kind. Commonly *Einraumbeute* users on the European continent use mushroom head spacers made of wood or metal, and inserted into holes in the frames. Instead, in the present case, pieces of bamboo were used, rounded with sandpaper at the exposed ends (Figs. 16 and 17). The protruding length of each frame-to-frame spacer was set to give 36 mm between frame centres. One of these spacers was placed at the top and bottom of each frame side bar, facing opposite ways on each side. Having two such spacers, i.e. one at the bottom too, appeared necessary because of the unusually deep frames. Plastic Hoffman type spacers, as used in my modified Lazutin hive, would also have worked, and then would have required only one at the top of each frame side bar. In addition to the frame-to-frame spacers, in order to avoid squashing bees when lifting or lowering a frame, side spacers were added at the bottoms of the side bars to hold them away from the hive walls (Fig. 15, lower spacer).

*Fig. 17. Frame with bamboo comb supports*

The idea for the bamboo comb supports came from an article by Thomas Radetzki on Mellifera Association's website.[21] My supports comprise bamboo dowel, sold as skewers, 3 or 4 mm in diameter and about 50 mm long. Their advantage over wiring the frames is that there is no interference with the comb in the heart of the brood nest, only at the edges. And as there was no intention of extracting honey by centrifugation, the additional comb stabilisation given by wiring the frames was unnecessary.

The *Einraumbeute* (golden hive) frame has an inner area of 1166 cm$^2$ whereas that of the National hive is 636 cm$^2$, or only 54% of the *Einraumbeute* comb area. Therefore only six *Einraumbeute* frames are needed to make up the comb area of a standard 11-frame National brood box. So 12 frames would represent a National double brood hive, leaving a further 10 Einraumbeute frames for honey stores, equivalent to about three 10-frame National supers. The already mentioned popular horizontal hive of de Layens has frame inside dimensions of 370 mm deep by 310 mm wide, a little shallower and wider than the *Einraumbeute* frame.[22] This gives a single frame internal area of 1147 cm$^2$, i.e. very close to that of the golden hive.

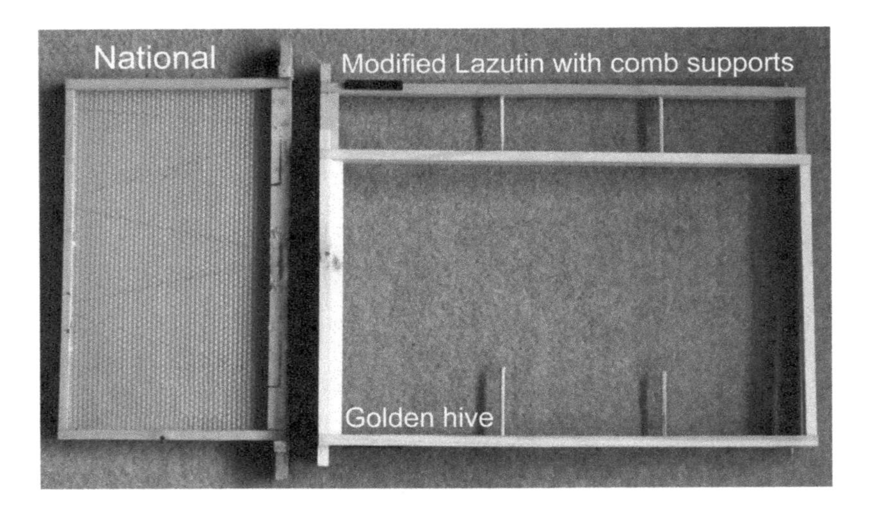


*Fig. 18. Comparison of National hive, Golden Hive and modified Lazutin Hive frame*

Although the frames in the modified golden hive described here have a 'V' cross-section to provide a guide for the bees on where to build comb, the precaution was taken to add wax starter strips to the bottom of the 'V' of the top-bars. This was achieved by using a template with surfaces at the appropriate angles in order to create a vertical starter-strip. The template was first soaked in cold water, placed against the inverted top-bar and melted beeswax poured along the template and top-bar.

*Fig. 19. Template positioned for waxing top-bars*

*Fig. 20. Frames with waxed top-bars, comb supports and spacers*

*Fig. 21. Waxed top-bars (detail).*

*Fig. 22. Follower board and foam seal off rear of chamber.*

One of the frames was used to make a follower board by nailing thin sheets of plywood on both sides while packing the resulting cavity with insulating material.

## Management

This little book does not give detailed instructions on management of the modified golden hive. For this, the reader is referred to the excellent book by Johannes Wirz and Norbert Poeplau who have very long experience of the golden hive.[23] Other useful books to consult on management of horizontal hives are those of Georges de Layens[24] and Fedor Lazutin.[25] Even the book of Les Crowder is worth consulting because, although it deals with the horizontal top-bar hive, i.e. no frames, its comb management guidance is very clear and richly illustrated by coloured diagrams.[26] Readers are also advised to learn all they can about bee biology and behaviour.

*Fig. 23. Running in a swarm*

In the rest of this book, I present a few highlights of running my modified golden hive over six seasons and give a few examples from other beekeepers using the hive. Mine was managed with minimal intervention and entirely without Varroa treatment or swarm control.

It was inspected by the local seasonal bee disease inspector in 2018, himself a fan of deeper frames, albeit 'fourteen by twelves', who commented on how calm the bees were.

Having made and installed the hive in the winter of 2014-2015, I populated it the following June. Eight frames with starter strips were placed at the entrance end of the hive. A division board closed off the unused part of the hive. This was adapted from the follower board by temporarily packing flexible plastic foam round the bee spaces of the follower board (Fig. 22). Folded hessian or another fabric could equally well have been used for this.

A swarm of about 1.5-2 kg was run in late in the evening. To minimise the risk of absconding due to the odour of new wood in the hive, a queen 'includer' was placed over the entrance for 24 hours (Fig. 25). As the swarm came to a bait hive situated only 1.6 km away from this hiving, and there was the risk that some field bees might return there, leafy branches were placed in front of the entrance in order to force reorientation (Fig. 26).

*Fig. 24. Swarm hived*

*Fig. 25. Queen includer on entrance*

*Fig. 26. Obstructing the hive exit*

Two weeks later there was already comb on the eight frames (Fig. 27). But on inspection there was brood on only one comb, and then only a few drone pupae (Fig. 28). A few scattered eggs were found on comb 2 counting from the entrance and in the entrance side of comb 3. Some cells had several eggs. The colony was quiet on the combs and apparently foraging normally. A queen was observed on comb 1. The hiving appears to have included a defective queen. In any case, the colony was not 'taking off' as expected.

*Fig. 27. Developing colony*

*Fig. 28. Brood on comb*

As it was July, therefore late in the season, and the colony was dwindling due to the lack of worker brood, adding more bees and a laying queen was considered the best way of remedying the situation so that the colony could become winter-ready. Fortunately, only a few days later, a good sized swarm turned up in the nearby town. The golden hive combs were again inspected, revealing that the brood situation had not improved. The queen, a fine-looking black specimen, was removed in a queen clip and placed by the entrance to check whether the colony may have had more than one queen. After about 10 minutes the bees were seen running around looking for the queen and taking to the air, eventually landing on the clip containing their queen. The way was then clear to unite the swarm with the colony using the newspaper method.

*Fig. 29. Colony after uniting with swarm*

*Fig. 30. Brood on comb*

The uniting went ahead without fighting and by four days later the colony population was visibly greater (compare Figs. 27 & 29). A fortnight after the uniting, a good sized patch of capped worker brood was observed on comb 2 (Fig. 30, arrows).

The combs were next inspected in September in order to estimate the weight of honey stores available for wintering. The result was only 2.8 kg. As my Warré hive colonies are usually wintered on 9 kg stores, it was obviously necessary to give the golden hive colony an emergency feed. This comprised 6 kg sugar as 2:1 syrup. It was presented in a bucket next to the follower board in the empty part of the hive. The syrup surface was covered with wine bottle corks and twigs were placed to enable the bees to climb in and out of the bucket. Sufficient corks are used to prevent them rolling and thereby drowning bees. This means having a few extra corks in a second layer on those that are floating. (Instead of wine corks, the traditional way is to put straw in the bucket.) During the autumn, plenty of ivy pollen was seen being brought in at the hive entrance. A mouse guard was fitted in October.

The following spring, the colony built up very rapidly, in fact more rapidly than frames were added to the nest side of the follower board, to the extent that comb was being built outside the nest. After moving the follower board backwards, soon comb 21 was being built right at the back of the hive (Fig. 31).

*Fig. 31. Comb 21 (rear).*

Although the frames were given prominent starter-strips, the speed of development combined with neglecting honey comb management, led to a certain amount of cross-comb, i.e. combs fixed to two top-bars. According to Wirz and Poeplau, there needs to be a at least some management of comb construction in the golden hive. They advise correcting irregularities with the hive tool, or even rotating wavy combs horizontally.[27] This obviously requires several hive openings during the main nectar flows, but as access is then needed to only part of the combs, the top-bar cloth has to be rolled back only partly and the front quilt left in place on the brood nest (Fig. 32).

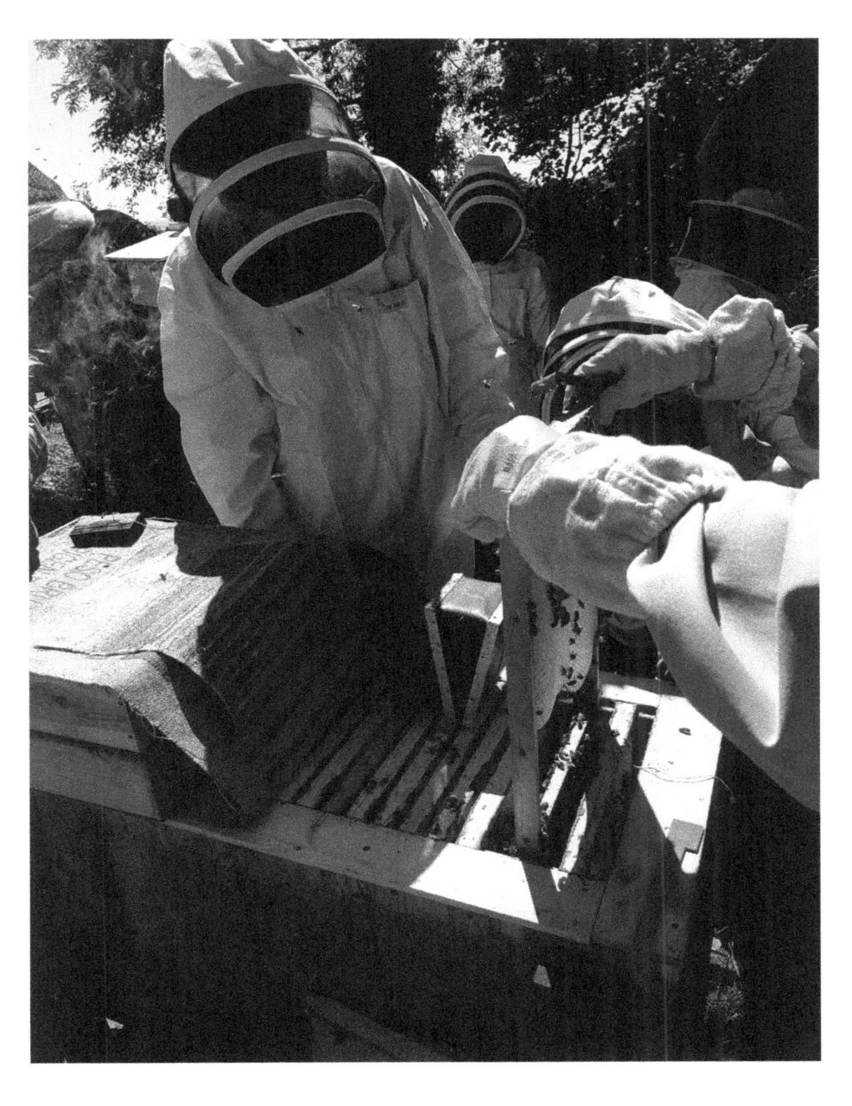

*Fig. 32. Inspecting the honey zone of the modified golden hive during a local beekeeping association meeting*

In September of the second season, it was possible to harvest 8 kg honey while leaving sufficient so that no winter feeding was necessary. To harvest the combs of honey, it is useful to have a lightweight closable transport box that will accommodate several frames with their lugs resting on battens in the box. This minimises the risk of triggering robbing or taking excessive numbers of wasps and bees home with you. Combs for harvest are swept free of bees with a large feather, or failing that, a bee brush, and placed in the transport box. The box could double as a bait hive or even a swarm collection box. In the these forms it needs an entrance and meshed vents, all closeable.

*Fig. 33. Comb ready for extraction*

Combs for extraction (Fig. 33) are placed over a rectangular food-grade plastic tub, sliced into the tub, crushed and drained through a sieve (Figs. 34 & 35). If required, the honey can be further strained through a jelly bag.

*Fig. 34. Cutting comb out of frame*

*Fig. 35. Draining crushed comb*

In addition to the 8 kg harvested in 2016, further harvests of 11 kg (2017), 37 kg (2018) and 24 kg (2019, 13 frames) were obtained. The average harvest over those four seasons was 20 kg. But in the summer of 2020, the queen failed, and the opportunity was taken to harvest the honey and clear out the old combs ready for a new colony the following season. The colony had survived untreated for five years. It may have swarmed or superseded its queen in that time, possibly more than once, but as intervention was minimal, no information is available on queen replacement frequency. The hive was repopulated with a swarm in June 2021.

## Other Modified Golden Hive Projects

As already mentioned, the modified golden hive of Matt Somerville and John Haverson has, in addition to the double walls and Warré style roof, a space under the frames accessible through a rear hatch (Figs. 36 & 37). Their project is shown in detail on a web page where downloadable plans can be found.[28]

*Fig. 36. Modified golden hive of John Haverson and Matt Somerville (front view)*

*Fig. 37. Modified golden hive of John Haverson and Matt Somerville showing rear removable inspection hatch with vents*

Matt Somerville also has a web page on his modified golden hive where he writes: "... from using this hive for the past few years, I have noticed that we simply don't have the same environment as Lazutin's wild Russia and the flow of nectar is not nearly as great. This has been noted in the press with a marked decline in yields over the last few decades, despite many beekeepers feeding sugar in Autumn and Spring. [...] This year I designed a smaller hive using 13 frames of the same frame dimensions but having a few new changes". His modified golden hive with 13 standard *Einraumbeute* frames and a follower board is presented on a page of its own together with many photos, and is available to order.[29] Although the hive has double walls, there is no insulating cavity between them. However, the walls are made of two layers of 27 mm cedar, a wood with inherently low thermal conductivity. It also has the Warré style quilt, a gap under the frames and removable base boards for cleaning and inspection.

Mark Tan's project is based on the aforementioned plans of Haverson and Somerville (Fig. 38).

*Fig. 39. Mark Tan's modified golden hives*

There also exists a modified *Einraumbeute* design that compromises between wooden and straw walls. It is offered for sale by Robert Friedrich (Germany) and comprises an outer box of wood lined with rye straw on the inside.[33]

David Bolton's hive has its round entrances about two-thirds of the way up the hive's long side (Fig. 39). He plans to slightly reduce the frame size while preserving the golden ratio so that his frames will fit in his extractor.

*Fig. 39. David Bolton's modified golden hive*

As well as projects based on wooden-walled modified golden hives, there are a few projects in Europe and the USA using walls comprising mats of packed straw, albeit with wood at the corners.[30] The technique, comprising pressing in a wooden frame long rye straw that has preferably been cut by hand, is illustrated in Matthias Thun's book, although not in its English edition.[31] The same method was used by Uwe Bodenschatz who led a workshop making a straw golden hive with Michael Thiele and associates in California (Fig. 40). Videos of the construction, as well as of the populated hive, are available.[32]

*Fig. 40. Michael Thiele's straw golden hive*

[1] Warré, É. (2007) Beekeeping for All, Northern Bee Books, Mytholmroyd. Free download: http://www.users.callnetuk.com/~heaf/beekeeping_for_all.pdf .

[2] http://www.dheaf.plus.com/framebeekeeping/oneboxhive.htm

[3] http://www.dheaf.plus.com/framebeekeeping/modified_golden_hive.htm

[4] https://www.mellifera.de/einraumbeute/.

[5] Wirz, J. & Poeplau, N. (2021) Keeping bees simply and respectfully – Apiculture with the Golden Hive. International Bee Research Association and Northern Bee Books, Mytholmroyd. Translation of Imkern mit der Einraumbeute, Pala Verlag, 2020. Page 47.

[6] Radetzki, T. (2003) Die neue Einraumbeute von Mellifera. Mellifera e.V. Fischermühle, Germany. PDF booklet download: https://silo.tips/download/die-neue-einraumbeute-von-mellifera. Page 2.

[7] Ibid.

[8] https://www.horizontalhive.com/

[9] https://www.thorne.co.uk/index.php?route=product/product&product_id=4850

[10] Dartington, R. (2008) Introduction to New Beekeeping with Dartington Hives. http://www.thorne.co.uk/image/data/Dartington/Dartington%20document/B2%20%20DLD%20INTRODUCTION%20SPRING%2008%20P%20CMP%20-%2010-1-14.pdf.

[11] https://www.omlet.co.uk/shop/beekeeping/beehaus/.

[12] https://www.draytonbeehive.com/

[13] https://www.thezesthive.com/

[14] Lazutin, F. (2009, 2013) Keeping Bees with a Smile – A vision and practice of natural apiculture. Deep Snow Press, Ithaca, N.Y. p. 110.

[15] Haverson, J. & Somerville, M. (2014) Plans for the Modified Golden Hive. http://www.users.callnetuk.com/~heaf/modified_golden_hive_construction.pdf.

[16] https://www.mellifera.de/blog/mellifera-einraumbeute/einraumbeute-bauanleitung.html

[17] https://www.mellifera.de/angebote/einraumbeute/erb/ausstattung.html .

[18] Wirz, J. & Poeplau, N. (2021) Keeping bees simply and respectfully – Apiculture with the Golden Hive. International Bee Research Association and Northern Bee Books, Mytholmroyd. Translation of Imkern mit der Einraumbeute, Pala Verlag, 2020.

[19] Warré, É. (2007) Beekeeping for All, Northern Bee Books, Mytholmroyd. Free download:http://www.users.callnetuk.com/~heaf/beekeeping_for_all.pdf . Page 53.

[20] https://beekindhives.uk/. email: beekindhives@gmail.com

[21] https://www.mellifera.de/blog/biene-mensch-natur-blog/stabilisierung-des-naturwabenbaus.html

[22] de Layens, G. (1891) Construction économique des ruches à cadres. https://gallica.bnf.fr/ark:/12148/bpt6k318382g/f14.item.texteImage/f1n36.pdf

[23] Wirz, J. & Poeplau, N. (2021) Keeping bees simply and respectfully – Apiculture with the Golden Hive. International Bee Research Association and Northern Bee Books, Mytholmroyd. Translation of Imkern mit der Einraumbeute, Pala Verlag, 2020.

[24] de Layens, G. (2017) Keeping Bees in Horizontal Hives – A Complete Guide to Apiculture. Deep Snow Press, Ithaca, N.Y.

[25] Lazutin, F. (2019, 213) Keeping Bees with a Smile – A vision and practice of natural apiculture. Deep Snow Press, Ithaca, N.Y.

[26] Crowder, L. & Harrell, H. (2012) Top-Bar Beekeeping – Organic Practices for Honeybee Health – Natural hive management for honey, beeswax and pollination. Chelsea Green Publishing, White River Junction, Vermont

[27] Wirz, J. & Poeplau, N. (2021) Keeping bees simply and respectfully – Apiculture with the Golden Hive. International Bee Research Association and Northern Bee Books, Mytholmroyd. Page 96.

[28] http://www.dheaf.plus.com/framebeekeeping/modified_golden_hive.htm.

[29] https://beekindhives.uk/the-modified-golden-hive/

[30] https://discourse.mellifera-netzwerk.de/t/erfahrungen-meinungen-zu-einraumbeute-mit-stroh-waenden/1215/36 and

[31] Thun, M. K. Die Biene, Haltung und Pflege. M. Thun-Verlag, 2015. Pages 82 & 83.

[32] https://www.youtube.com/watch?v=fq5ZTtNfonM and https://www.youtube.com/watch?v=zlPBkfB5vCk and

[33] http://strohbeuten.de/stroh-einraumbeute-neu/